"*Models of Reality*" for Static, Nuclei and Atoms

J. Alden Erikson

First published by AuthorHouse 06/16/05

ISBN: 1-4208-5250-7 (sc)

Library of Congress Control Number: 2005905074

Printed in the United States of America
Bloomington, Indiana

This book is printed on acid-free paper.

authorHOUSE

1663 Liberty Drive
Bloomington, Indiana 47403
(800) 839-8640
www.authorhouse.com

Figures

Tables

Contents

Preface

Considering the very slow rate of improvement in understanding of more useful nuclear and atomic models, a proposition is suggested for nuclear chemists and nuclear physicists to consider to advance the understanding of these challenging, minute entities of matter. Theoretical obstacles learned in college, the necessity to create a satisfactory life and earn a living delayed or made the pursuit of this new approach to knowledge very slow.

Now it appears that a few of the elements would like to reveal what their composition is and what it means. This paper encompasses my conjectures partly in chronological order and mostly as it appeared new conclusions were warranted.. There are so many stable isotopes and so many more unstable ones and chemistry that goes with all of them that no one man should try to explain all the ramifications of the physical models. Albert Einstein predicted that "models of reality" would be found that would make nuclear physics and chemistry more understandable. Maybe the information presented here will lead to that better understanding in ways that have eluded wave mechanics, quantum mechanics, atomic orbitals, molecular orbitals, chemistry and physics for the past 80 years at less.

It would be nice if an old man could contribute something to science and even nicer if a direct descendent of Lief Erickson could make a discovery in chemistry equivalent to his discovery of America. about 1100 years ago.

J. Alden Erikson,

Introduction

Way back in 1913 Niels Bohr provided the understanding needed to get from atoms to atomic spectra. The atomistic interpretation of the elements in the periodic table soon followed. Atomic structure was the result of electrons in discrete locations in the atom. But the Bohr theory was not rigorous and consistent. It did not explain why each consecutive electron takes one and only one position in the atom. Actually there were more problems than one might imagine that arose with time. The net result of all the unsolved difficulties required a new theory, introduced in 1926, called "wave mechanics" based entirely on observed phenomena, discrete energy states, and "*transition probabilities*."

In 1933 Albert Einstein appreciated the successes of both classical mechanics and the statistical interpretation of quantum theory/or wave mechanics, but he believed that, *"Some day there would be a* ***model of reality*** *which would represent events themselves and not merely the probability of their occurrence*." I believe scientists in the 21st century may be able to use some of what is presented here to permit improved understanding of atoms and nuclei progressing from classical science to modern physics and back again, understandably.

By 1947, Richtmyer and Kennard wrote in their book, *Introduction to Atomic Physics,* that individual electrons in a many electron atom move in a charge cloud created from the positive nucleus and all the other electrons. Furthermore, according to wave mechanics, they do not occupy definite positions but only have a probability of being found in fixed positions. When you are a sophomore studying for a B.S. in Chemistry that is perfectly reasonable and you believe it. Your teachers know! What you learn is all true!

In 1963 Blatt and Weisskopf in their book *Theoretical Nuclear Physics* suggest that the great success of the shell model of the nucleus is because the Pauli exclusion principle prevents the transfer of momentum and energy between nucleons if they are all in their ground states, their lowest-possible, energy-states. Since the shell model of the nucleus is based on particles, useful models should be possible and all you have to do is find the basic models of Einstein's*"**reality**."*

Wassily W. Poppe and I presented a paper at the September 1967, Chicago American Chemical Society meeting, *Nuclear Models of the First Nine Elements,* in which we presented carbon nucleus as an

icosahedron. Later Lynus Pauling was kind enough to read the paper and he said we would probably have a hard time convincing others that what we had was right. I also understand Pauling spent two years, much later and not long before his death, trying to correlate nuclear particle models with known data, unsuccessfully. He taught that carbon-12 atom had a tetrahedral structure. It will be shown here that carbon may be better represented by an icosahedron.

Let's update the problems. *Time* Magazine, December 28, 1992, says the upshot of quantum physics is inexplicability and quotes the great physicist Richard Feynman as prefacing a lecture by telling the audience not to worry about not understanding it. "I think I can safely say that no one understands quantum mechanics." W.S.C. Williams, in his 1991 book *Nuclear and Particle Physics,* on p, 344 of 385 pages said, "Quantum mechanics. That is a very successful system but nobody understands why it is the way it is." That is the problem I have lived with the larger part of my spare time in a working lifetime. I wanted to understand how the nucleus related to the atom. Needless to say my understanding does not progress rapidly.

An attempt was made to use the carbon-icosahedron nucleus to develop the nuclei of some additional elements and a paper was presented by me at the ACS Central Regional Meeting in Pittsburgh in October 1993. But atomic nuclei are mostly spherical and should fit the nucleon binding energy curve provided by Williams on page 55. All of them didn't fit so well so I went back home and retirement and looked for better models.

Developmental work.

Five icosahedrons were made from poster paper and an attempt was made to put them together, pack them, to make nuclei. Suddenly it became apparent, an icosahedron with more or less equally distributed neutrons and protons was comprised of two hydrogen-3 nuclei and two helium-3 nuclei. If you choose one face, that face, along with three others, point toward the corners of a tetrahedron, *like electrons in a carbon atom do.* That is the first, new, observed, structural relationship of *"**reality**"* between nuclei and atoms in this paper! Can you build from this beginning? Can you work backward to other isotopes, or from carbon to boron for example, or up to nitrogen, oxygen to neon? Williams, on page 123, says the empirical mass formula applies only to elements with a mass greater than 20 so maybe that is why nuclear structure changes the way ir does and is different "above $A = 20$". Let's go on.

The new theory should be comprehended easily so that work can be carried on to make the nuclei and atoms more understandable. That is supposed to be one of *Physics World*'s ten most important goals for the 21st century, according to an excerpt in C.&E. News' *Newscripts*, Feb. 7, 2000. Then it may fit like Jodie Foster's use of "Occam's razor," a basic, scientific principle/precept from about the year 1345, referred to in the movie *Contact*: "All things being equal, the simpler explanation tends to be the right one!" Is there a simpler explanation? Are things equal? They will seem to be.

1. The First Key Element, Carbon 12

Look in your old solid geometry book and you might find how to make an icosahedron from heavy paper. You will find, when it is made, that it has 12 vertices and 20 faces. If you are lucky, you will find that the radius of an icosahedron is less than the distance between any two adjacent vertices on its surface. Now imagine that each vertex is the center of a nucleon, a proton or a neutron, and the distance between the centers is one nucleon diameter so the nucleons just touch one another on the surface of this regular polyhedron. But, in actuality, there is insufficient room to have a nucleon at the center without breaking or extending one or more bonds on the surface of the icosahedron. Note that it is a close-packed structure and that it can be enclosed by a sphere. The model of greatest interest here is considered a ground state nucleus of the atom.That makes it unique among possible geometric structures with 12 nucleons, as depicted in Fig. 1, and it might have a lower binding energy per nucleon than helium-4, as shown by Williams on p. 55.

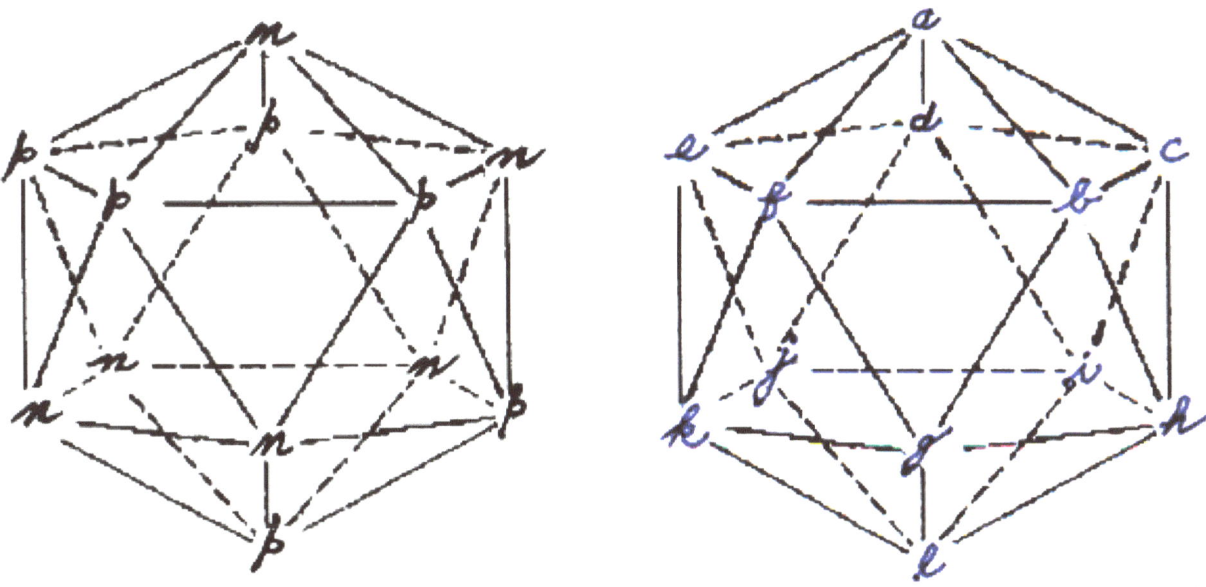

Fig. 1. Carbon with p's and n's specified. **Fig. 2. Vertices Identified by letters in Carbon.**

Now imagine the nucleons are located in their positions of greatest probability which might be defined by qantum mechanics for a moment in time. If you try to arrange the protons and neutrons so they will be equally distributed, as shown in Fig. 1, you will find that it is formed from two H-3 and two He-3 nuclei. Again choose a face, and that face along with three others, will point toward the four corners of a tetrahedron. That should remind everyone of the typical tetrahedral carbon atom and it may form a new basis for carbon chemistry.

The presence of the two helium nuclei seems to dictate that they each will have one electron above and one below the planes of their faces. That means this carbon can have 4 extra-nuclear electrons for *valence* purposes. One of the cliches of physics says you cannot confine electrons inside a nucleus, so I ignore that limitation to make the chemistry come out right. Electron size, 1/1836th of the mass of a nucleon, makes them small enough to fit through the interstices of the carbon nucleus making them very difficult to find inside the nucleus. Maybe they just neutralize two protons thereby making it appear carbon nucleus is formed from four H-3's. Now hybridization of the carbon electrons is completely understandable. But back to the fundamental carbon model. Fig. 2. shows a systematic lettering of the interstices in carbon icosahedron, the foundation for future chosen models. Interstices d, e, f, b, h and l are the proton locations and a, c, i, j, k, and g are for the neutrons. Protons e and h are opposite each other. Neutrons at c and h are opposite each other. I have assumed these positions portray the K shell of carbon nucleons. The other four protons are opposite neutrons. One can see distinct pairs of protons and/or neutrons anticipating the electron pairs designated 1*s*, 2*s* and 2*p* in carbon atom by the 1*s*, 2*s* and 2*p* nucleon levels.

Start there and you should be able to add a neutron to an a-e-f-He-3 face, to make a He-4 tetrahedron there and get carbon 13. Add a neutron to the other tetrahedrally directed He-3 face to make another He-4 tetrahedron at the new location. That would provide a possible *C-14.* A question arises as to whether the neutrons will be found as they are in neon, to be described shortly, or in these locations as conjectured.

To be more specific, the twenty faces were numbered as follows.1 = face afb, 2 = face abc, 3 = acd, 4 = ade and 5 = aef. Face 6 was designated below face 1 and face 6 = face fbg. Proceeding to the right, one continues to number the ten equilateral triangles forming the equator of the icosahedron up to face 15. Then face 16 is numbered below face 15 and equals face kjl. Face 17 is below face 7, face 18 is below face 9, face 19 is below face 11 and face 20 is below face 13.

2. Model of Neon-20

This brings us to the next *chosen* model, Neon-20. It is one of Williams' first 20 elements that are different than the higher atomic weight elements, requiring a change in scale for their growth rate as the binding energy changes (p. 55). Add protons to faces 2, 15, 10 and 20 of the carbon-core icosahedron. Add neutrons to faces 5, 7, 18 and 12. That gives $_{10}Ne^{20}$. The centers of the new shell of nucleons form a perfect cube. Neon coincidentally crystallizes with a cubic structure. The model corresponds to eight alpha particles sharing the corners of their bases and, just maybe, that means the nuclear structure determines the chemistry of the atom making neon a noble gas formed from 8 helium atoms. Maybe too the four new protons determine the levels of the first 4 electrons in the **L** shell of neon. Possibly, because only the available helium-3-type faces 4 and 17 are opposite one another to accept neutrons to form stable isotopes neon-21 and neon-22, that is where the necessary neutrons get added to make these stable isotopes. Possibly too, helium-3-faces 1, 6, 8 and 14 could accept neutrons and provide the unstable isotopes *23*, two *24*'s and *25.*

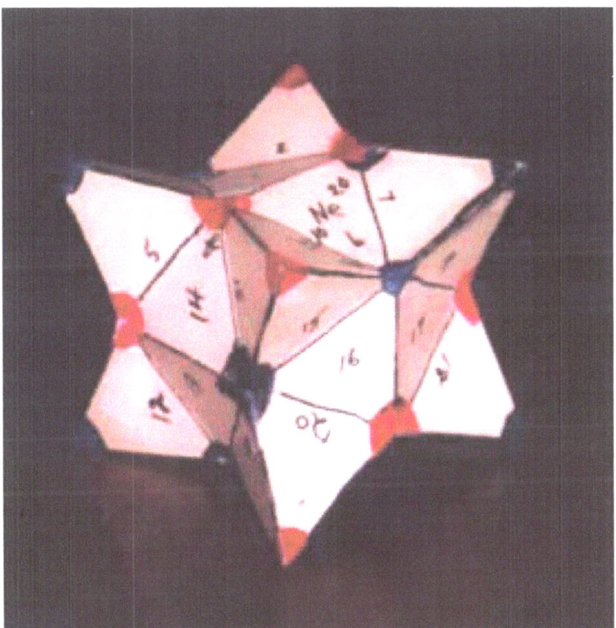

Fig. 3. Model for Neon 20

It is unfortunate but one probably will have to make the models to see the possibilities most effectively. Neon-20 was formed here by a different method of construction from what will be described for many higher atomic wright elements.

Going back to *C-14*, it seems possible someone would have discovered hydrogen-4. It would appear in isotopes, as proposed here, when new H-4 tetrahedrons are formed at tetrahedrally directed H-3 faces by adding neutrons. That is a way to get two *C-15*'s easily. It seems to be a possibility in unstable, carbon-isotope-development at least. If you keep both those added neutrons you have *C-16* but if you keep them both and remove a neutron alternately from one of the alpha-particle components of *C-16*, you can get models for two other *C-15* isotopes. Could these observations be an explanation for the ever increasing number of isotopes associated with higher atomic weight elements. It seems possible here at least.. Those are ways to proceed from C-12 to C-13 to *almost- stable C-14*, possible *C-15's* and *C-16*. Imagination and model building is fun, in fact or in the minds eye.

On the other hand, remove a *proton* from a He-3 face on *C-12* and you get stable B-11. Remove an adjacent *neutron* from B-11 and you get B-10. Wonderfully those partial icosahedrons have only three complete valence faces corresponding to the three valence faces in carbon nucleus and accounts for the valence of 3 in boron. Can that be true? Obvious possible differences in isotope chemistry seem to be possible if one examines the B-10 and B-11 models and their different faces. It will take a lot of model construction work and imaginative deduction about proton-neutron optimum distributions to correlate structures to chemistry, but they seem very possible today.

Now remove a proton adjacent to the hole formed from removing a proton and neutron from carbon icosahedron and get Be-9 with a face missing from an icosahedron as shown in Fig.4.

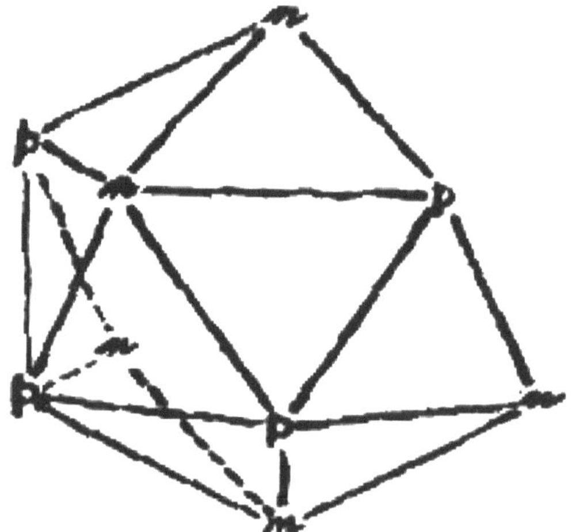

Fig. 4. $_4Be^9$ for addition to a $_2He^3$ face.

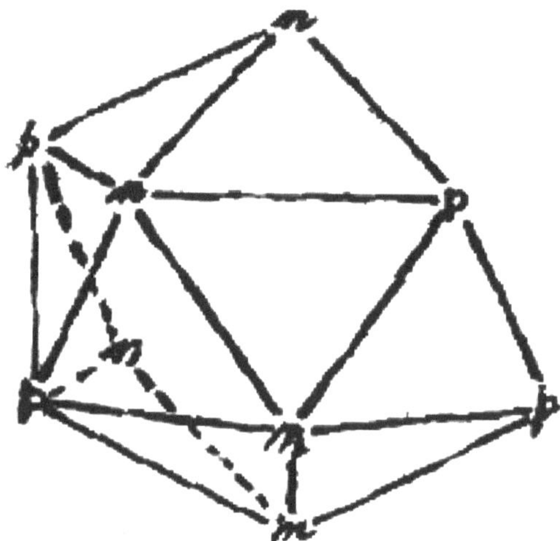

Fig. 5. $_4Be^9$ for addition to a $_1He^3$ face.

Apparently Be-9 can be made two ways, with one He-3 and two H-3 nuclei, if not before reaction with carbon-icosahedron at least after depending on whether addition is to a He-3 face or to a H-3 face. I feel it is important that every face on the stable models have one or two protons.

3. Key Model Titanium-48

Two Fig. 4-type Be-9's and two Fig. 5-type Be-9's are added to carbon icosahedron to get to Ti-48.The part of an icosahedron shown in Fig. 4 is identical to that part of a carbon-icosahedron and condensation with a He-3 carbon-core face forms a fused, carbon-type icosahedron with 6 protons. Maybe each of the He-3 faces farthest from the center of the model might accept a neutron to give Ti-49 and Ti-50, two of the stable, higher-atomic-mass, titanium isotopes.

In Fig. 5 is shown what might be the model structure of the Be-9 fragment after one of the Be-9's has reacted with one of the two H-3 faces on carbon nucleus. The part of an icosahedron shown in Fig. 5 is new and with the carbon-core H-3 face it fuses to, it forms a *boron-12*-type icosahedron with only 5 protons. That is what is dictated to the model builder. Their outermost faces are H-3's and maybe each can lose a neutron to provide stable, titanium isotopes 47 and 46. This is why this paper is said to find **Key** element-isotopes, the **Key** isotopes around which the higher and lower stable and unstable, individual, element-isotopes, can be formed. One face on each boron-12-type icosahedron has no proton on it and maybe this begins to explain why higher molecular weight nuclei become radioactive, have more neutrons and ultimately have shorter half lives

Now, looking at the different levels of the **Ti-48** model made with individual particles and looking at it as a shell model of a nucleus, it can be seen, as a first approximation, that there are 6 protons and 6 neutrons on the surface of the core carbon-icosahedron and it is suggested now that this icosahedron might be called the **K** nuclear shell with two $1s_{1/2}$, and four $2p_{3/2}$ protons or neutrons..

The new shell of four formed-icosahedrons, made from the addition of two Fig-4 and two Fig-5 beryllium-9-type icosahedrons to the carbon-core, together can be thought of as the **L** nuclear shell. Counting the protons in each **L** sublevel gives $[(2p_{1/2})^2],[(3d_{5/2})^6,(2s_{1/2})^2],[(3d_{3/2})^4,(4f_{7/2})^2]$ equal to a total number of [2], [8] and [6] protons in each successive sublevel. This agrees in principle with Finkelnburg's 1964 designations for the protons forming Tin-50 on p. 277 in his book *Structure of Matter*. It is intriguing that the outermost faces of the titanium model are like carbon's, two H-3 and two He-3 faces. Therefore one should not be surprised that titanium prefers to make tetrahedral compounds and forms these covalent compounds much like carbon does!. Some way, someday, I predict, the twelve tetrahedrally-directed faces of those four *L* shell icosahedrons are going to explain the other valences and coordination chemistry displayed by titanium and noticeable differences between the chemistry of isotopes if the neutrons are lost or gained as I suggested. Scientists must get the electrons involved correctly in order to understand chemistry and physics of nuclei completely. Maybe I will do it some day.

Going back to chemistry, Be-9 is usually divalent when forming compounds. Four of them used on **Ti-48** might make it octavalent and that too is found in titanium chemistry. Two **Ti-48** nucleon-models are shown in **Figs. 6** and 7. Models get more difficult to build. One must build logically on the twelve tetrahedrally-directed faces on **Ti-48** using four faces at a time to get to $_{36}\mathbf{Kr}^{84}$.by forming four, appropriate, additional icosahedrons resulting in what is called here Shell 3, or maybe it is equivalent to quantum level M as shown in **Table I**..

Fig. 6 Titanium-48 Force-line Model.

Fig. 7 Ti-48 Made with Spherical Particles

Table I. Nucleon-Additions to Form Key-Element, Icosahedron-Based Models.

Element-*A*	Protons P	Neutrons N	Protons Added	Neutrons Added	Total No. Nucleons	of Completed Added Shells
Carbon-12	6	6	—	—	—	1 = K
Titanium-48	22	26	16	20	36	2 = L
Krypton-84	36	48	14	22	36	3 = M
Tin-120	50	70	14	22	36	4 = N
Gadolinium-156	64	92	14	22	36	5 = O
Osmium-192	76	116	12	24	36	6 = P
***Radium*-228**	88	140	12	24	36	7 = Q

Some Observations about Table I

Kr-84. The isotope table showed it takes 14 protons and 22 neutrons to get from **Ti-48** to **Kr-84.** That means two of the partial icosahedrons used with two 4-beryllium-9's must be two 3-lithium-9's providing two fewer protons. Looking at the lithium isotopes in the isotope table showed there are three, unstable lithium-9's, which might be like those in **Fig. 8.**

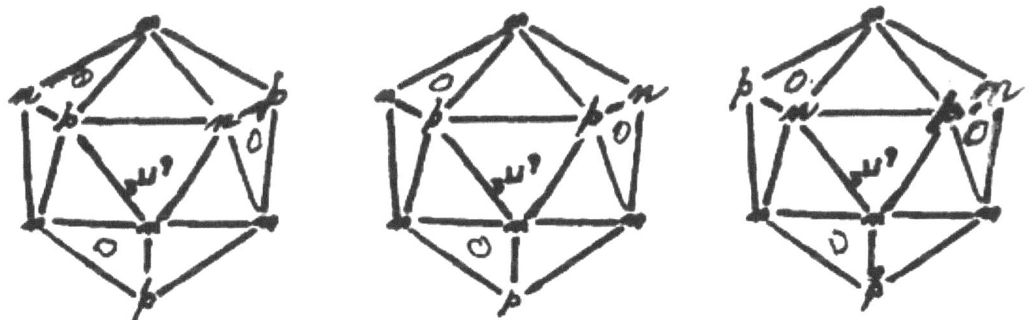

Fig. 8. Three Possible, Unstable, Partial-Icosahefron, Lithium-9 Models.

The Fig. 8 models look like some isomeric and stereoisomeric structures in organic chemistry formed from three H-3 nuclei marked by circles. The first and third figures seemed possible and might make it easier to lose a neutron or two once it becomes part of a nucleus. The middle model may account for the longest half life isotope. Without specifying individual choices, two of these Li-9 models would be used along with two beryllium-9's to get to **Kr-84.** Be-9's would go on H-3 faces to form B-12 icosahedrons with 5 protons. The Li-9's added to He-3 faces would also form B-12 icosahedrons. Could the four new, boron-type icosahedrons lose one, two, four or six neutrons to provide the four, stable, lower-mass-number isotopes 78, 80, 82 and 83? Or can **Kr-84** accept two more neutrons to form stable Kr-86? Unstable boron-13's are known and **Kr-84** with four boron-12's would seem to be able to accept neutrons on each of two of them to make the higher mass number isotope. It is suggested that the four new icosahedrons be considered Shell 3 or level M. A total of four icosahedrons were formed, one for each L-level icosahedron in **Ti-48**, to maintain a uniformity of structure.

Key isotope **Tin-120** is the most abundant of **10** stable isotopes including 112, 114, 115, 116,. 117, 118, 119, **120**, 122, and 124. Isotope**-120** was chosen to be made by the addition of 14 protons and 22 neutrons taken 9 at a time to form four more icosahedrons on two He-3 faces and two H-3 faces at the L level of Kr-84 forming Shell-4 or new level N. It is not clear why tin-120 has more isotopes than krypton bot once again four boron-12's would seem to be able to lose 8 neutrons to give the lower mass number isotopes or gain two or four to get to tins -122 and 124.It may be even more significant that the numbers lost or added are 8 or 4 as might be expected from the icosahedral models. There seems to be a fundamental relationship between the models and required, stable isotopes often.

Gadolinium-156 is the next key element. It has 7 stable isotopes. There are no more He-3 faces left on Shell 3 of **Ti-48**, only four H-3 faces. Maybe addition of two Be-9 and two Li-9 partial-icosahedrons to these faces would form two new B-12 icosahedrons and two Be-12 icosahedrons. Maybe two neutrons are added somewhere else in the nucleus as Li-8's are used?

Elemental Be-12 is not listed in the isotope table but as a model builder I think it is needed so it or they will lose neutrons to provide gadolinium isotopes 155, 154 and 152 as icosahedrons Be-11 and B-10 appear. The addition of 1,2 or 4 neutrons to get Gd's other stable isotopes-157, 158 and 160 by addition to the four, available, Shell 5, B-12-type icosahedrons is not so obvious but we have completed what could correspond to Level O. Gadolinium is spherically symmetric meaning it could be enclosed neatly by a sphere. Maybe its symmetry is the reason it is the only isotope appearing among the **key** element-isotopes that is not ranked first among the individual **Key**-element's other stable isotopes.

Fig. 9. Gadolinium-156's

To help the reader, **Fig. 9** shows a picture of a force-line model balanced on three of its icosahedrons that are attached to a **Ti-48** Shell 3 icosahedron. The other picture shows the same model balanced on a roll of masking tape to view it from a different aspect. It appears nuclei might seem fairly porous to particles the size of electrons. A model made with 192 little spherical nucleons would use up more of the space in the nucleus but icosahedron-type models are much easier to make and interpret.

Osmium-192 is the next, **Key,** icosahedron-based model and is the last one of **7** stable isotopes including 184, 186, 187, 188, 189, 190, and **192**. **Table I** shows the new Shell, 3, is made using only 12 protons and 24 neutrons, 9 nucleons at a time, to form four, new icosahedrons.. This means four, Li-9-type partial-icosahedrons can be used and just account for all the needed nucleons; (4 x 3)protons + (4 x 6)neutrons = (12 + 24)nucleons = 36 nucleons as designed.

Finally, what looked like the last **Key** isotope, though radioactive or unstable, is **radium-228,** the isotope with the second longest half-life among the radium isotopes listed. It would be formed the way osmium is, requiring 12 protons and 24 neutrons. Somehow the higher number of neutrons now present in the element-isotopes makes them radioactive. Therefore what gadolinium and osmium and radium models really look like will have to wait at least until my ophthalmologist and optometrists help me to see well again. (A capsulotomy is needed..) The ground state, structural relationships still fit the icosahedron model thesis. This observation made me wonder about all the intermediate icosahedron models, if completed one icosahedron at a time, and how they would fit in the scheme of things. There seemed to be a tremendous correlation here between the periodic table and the new models. The 7 successive shell levels of icosahedrons that seem to coincide with the principle quantum numbers for the electrons need elaboration.

Therefore **Table II** was deduced from the **Key** ***A***, ***atomic masses***, derived using the equation (4,1), shown in the next section, and were used along with the chemical isotope Table in my *1990 CRC Press, Handbook of Chemistry and Physics*, to find the elements with those **Key** ***atomic mass*** numbers, which in turn showed the ratio of protons and neutrons needed in the completed, icosahedron-based models in both **Tables I** and **II**,..

4. The Individual, Completed-Icosahedron Models.

The proposed arithmetic equation is:

$$\text{Isotope Mass Number } A = 12 + 9I \qquad \textbf{(4,1)}$$

where 12 equals the core, carbon-icosahedron's nucleon number and ***I*** equals the number of partial icosahedrons formed on it, and subsequent models, to make completed elements. Unique mass number-isotopes were defined but the element name was obtained from the isotope table of the elements and the proton and neutron numbers came from those particular element-isotopes because they were among the stable isotopes at the equation-defined masses. Table II is an introduction to the consequences of building individual, icosahedron-based, nuclear models. It seems probable that they all will be constructed in some systematic, compact fashion to the tetrahedral faces of icosahedrons as described under **Table I**.

Table II. A Table of Completed, Icosahedron-Sized Element-Isotopes.

$_{p}$Element[4] Icosahedrons (***I***)	Core + Added	Total No. of Neutrons (N)	Comments, (Part. Icos. Added)	Rank, position
$_{6}C^{12}$	(C=1)+0 = **1**	6	99% of 2 stable isotopes.	1, 1st
($_{10}Ne^{20}$)	(1 + 8 nucleons)	10	90% of 3 stable isotopes.	1, 1st
($_{14}Si^{30}$)	1+2 = **3**?	16	3% of 3 stable isotopes.	3, 3rd
($_{19}K^{39}$)	1+3 = **4**	20	93% of 3 stable isotopes. ($_{5}B^{9}$)	1, 1st
$_{22}Ti^{48}$	(C=1)+4 = **5**	26	74% of 5 stable isotopes. (Li)	1, 3rd
$_{26}Fe^{57}$	1+5 = **6**	31	2% of 4 stable isotopes. (Be)	3, 3rd
$_{30}Zn^{66}$	1+6 = **7**	36	28% of 5 stable isotopes. (Be)	2, 2nd
$_{33}As^{75}$	1+7 = **8**	42	100%. Only 1 stable isotope. (Li)	1

Table 1I continued

$_{36}$Kr^{84}	(Ti=5)+4 = **9**	48	57% of 6 stable isotopes. (Li)	1, 5th
$_{41}$Nb^{93}	1+9 = **10**	52	100%. Only 1 stable isotope (B)	1
$_{44}$Ru^{102}	1+10 = **11**	58	31% of 7 stable isotopes. (Li)	1, 6th
$_{48}$Cd^{111}	1+11 = **12**	63	13% of 8 stable isotopes. (Be)	3, 4th
$_{50}$Sn^{120}	(Kr=9)+4 = **13**	70	33% of 10 stable isotopes. (He)	1, 8th
$_{54}$Xe^{129}	1+13 = **14**	75	26% of 9 stable isotopes. (Be)	2, 4th
$_{56}$Ba^{138}	1+14 = **15**	82	71% of 7 stable isotopes. (He)	1, 7th
$_{62}$Sm^{147}	1+15 = **16**	85	15% of 7 stable isotopes. (C)	3, 2nd
$_{64}$Gd^{156}	(Sn=13)+4 = **17**	92	20% of 7 stable isotopes. (He)	3, 4th
$_{67}$Ho^{165}	1+17 = **18**	98	100%. Only 1 stable isotope. (Li)	1
$_{70}$Yb^{174}	1+18 = **19**	104	32% of 7 stable isotopes. (Li)	1, 6th
$_{74}$W^{183}	1+19 = **20**	109	14% of 5 stable isotopes. (Be)	4, 3rd
$_{76}$Os^{192}	(Gd=17)+4 = **21**	116	41% of 7 stable isotopes. (He)	1, 7th
$_{80}$Hg^{201}	1+21 = **22**	121	13% of 7 stable isotopes. (Be)	4, 5th
<$_{84}$Po^{210}>	1+22 = **23**	126	*A* listed at 210. (Be)	-, 1st
<$_{86}$Rn^{219}>	1+23 = **24**	133	*A* listed at 222. (He)	
<$_{88}$Ra^{228}>	(Os=21)+4 = **25**	140	*A* listed at 226. (He)	
<$_{92}$U^{237}>	1+25 = **26**	144	*A* listed at 238. (B)	

In Table II, Completed Shell Key Isotopes are shown in **bold red** print. < > means the element is unstable. ( ) in the elements column means the carbon-core icosahedron has less than 4 icosahedrons attached to it. The Rank column tells the percent abundance of the isotope among the number of stable isotopes. Position tells where it falls among its stable isotopes. Neon-20 and Silicon-30 seem to be special cases. Ne-20 was described earlier and Si-30 has a question mark as to the number of icosahedrons attached to it, if any. Under comments, in parentheses, is listed the last partial-icosahedron added

Of the 26 elements listed, six of seven **red, completed-shell,** icosahedron-based, **Key** elements are stable and five rank number 1 among a total of 37 of all their stable isotopes. For the 19 other chosen elements, six icosahedron-based nuclei rank number 1 from a total of 56 of all their stable isotopes.

Therefore equation **4,1** selects 11 elements with a ranking of 1 among 26 elements and a total of 93 stable isotopes. There are two element-isotopes ranked 2nd, four are ranked 3rd and two are ranked 4th. The last four elements are not stable. These 26 selected element-nuclei are formed as described or from individual, completed-icosahedrons. All told there are 115 stable element-isotopes reported in the table under **Comments**. "Quantum mechanics" says elements with completed shells have more isotopes and greater abundance. A marvelous conclusion. The icosahedron-based element-isotopes found here seem to confirm that observation.

So chemistry and physics are coming together. Do the energy levels of the protons in the nucleus determine the energy levels of the electrons in the atoms? Are Shells as described here and Quantum, Levels comparable? They seem to be as seen in "Magic Numbers".

1. **Nuclear "magic" numbers**, according to Williams, are when the number of protons and/or neutrons, taken separately, equals 6, 20, 28, 50, 82 or 126. Of those, proton magic numbers 6 and 50 would appear as stable isotopes of the **Key** elements carbon and tin. Neutron magic numbers 20, 28, 82 and 126 would be found in isotopes of the **Key** elements potassium, titanium, barium and polonium. Adding a nucleon or taking one away from completed structures is said to be part of quantum mechanics related to magic number elements and that gives them their higher abundance of isotopes and more frequent occurrence in nature. That too agrees with what can be found here. It appears like *magic* when you see it for the first time..

2. **Atomic, "magic" numbers** are those whose ***proton numbers*** are **Z** = 10, 18, 36, 54, 80 and 86. Some of the atomic mass elements, below Ti-48 at least, may have fewer than 4 attached icosahedrons or be made differently. Neon-20, probably is built differently, but has Z = 10. Argon-40 could be C-1 plus three Be-9 partial-icosahedrons plus a neutron on the 4th tetrahedral face of the C-12 core to get $_{18}Ar^{40}$. The proton number equals a total of 6 + (3x4 = 12) = 18 for argon. Anyway, with these two element-isotopes all those proton magic numbers can be accounted for in neon, argon, krypton, xenon, mercury and *radon*, the last four already listed in the table. Model building can be fun. In the beginning I had neon-20 looking to me like a carbon-12 fused to a beryllium-9 to form neon-21 but that model didn't fit Table II as well as the neon proposed earlier. But now we need to look at how the elements are formed one icosahedron at a time because adding protons 14 at a time in Shells 3, 4 and 5 has to be different than making Shell 2 with 16 protons per four icosahedrons or Shells 6 and 7 with only 12 protons per Shell.

5. Possible Individual Icosahedron Growth Patterns.

Isotope Table numbers show the following growth patterns starting with carbon**:**

6-Carbon-12 icosahedron is Shell 1.

(10-Neon-20) may just be carbon-12 with 4 protons and 4 neutrons added to it.. No new icosahedron was formed but the nucleus is probably made from eight, helium-*atom* nuclei.

14-Silicon-30 would appear to have two Be-9 partial-icosahedrons added to carbon.

19-Potassium-39, surprisingly, has, possibly, a B-9 partial icosahedron added to Si-30**.**

22-Titanium-48 seems to have added a Li-9 to complete Shell 2. The atomistic interpretation of the elements implies electrons retain all their positions in elements as one proceeds through the periodic table. Do electrons fit with the locations of the protons in the nuclei?

26-Iron-57 adds a Be-9 partial icosahedron to one of the three kinds of Shell 2 icosahedrons. Does it add to a Li-9, Be-9 or the B-9 precursor partial-icosahedron?. I think I will keep in mind that iron is magnetic and the answer to why should come some day in the future. Maybe **Key** isotope **Ti-48** rearranges to the **Key** structure described earlier in Section 3. It is gradually becoming apparent why elements are so different from one another.

30-Zinc-66 might add the usual Be-9 to iron-57's Shell 2.

33-Arsenic-75 adds a Li-9. It is the only stable isotope. Why is arsenic so special?.

36-Krypton-84 forms as arsenic adds a Li-9 providing one icosahedron on each of the Shell-2 icosahedrons. The miracle is this makes krypton an inert gas. I wanted nuclear structure to determine chemistry and atomic structure. Now protons and neutrons have to be rearranged again to make the krypton nucleus like I thought it might be in the **Key** element **Table** I.

41-Niobium-93 adds B-9 partial-icosahedron to one of titanium's Shell 2 icosahedrons. It is the only stable niobium isotope and that too probably needs an explanation some day as to why.

44-Ruthenium-102 calls for a Li-9 partial-icosahedron added to a second Shell 2 icosahedron.

48-Cadmium-111 takes a Be-9 partial- icosahedron. I chose to put it on another Shell 2 icosahedron.

50-Tin-120 takes a He-9-type partial-icosahedron added to the fourth Shell-2 icosahedron. That, He-9, is another surprise thrown upon the theory. Maybe nucleons are added a different way between the **Key** elements and then rearrange to provide the most abundant, **Key**-isotope-structure. I had quit once inrending to let others decide how to build the models after Ti-48, with their more scientific approaches to solving problems. They just have to make themselves find understandable solutions. Then I realized

I had to make a stab at it myself and explored what was happening. 50-Tin-120 has to have the specified nucleon composition so a better explanation of its structure would be welcomed.

54-Xenon-129 is formed with the addition of a Be-9.partial-icosahedron to get to that inert gas.

56-Barium-138 requires a He-9 partial icosahedron.

62-Samarium-147 says it needs a carbon-9 partial icosahedron. That is a unique addition.

64-Gadolinium-156 calls for another He-9 partial icosahedron. That completes Shell 5.

And it probably explains why the chosen isotopes are not all ranked number 1 among their stable of isotopes. Nuclear chemistry is not quite so simple as it started out to be building icosahedrons.

67-Holmium-165 needs a Li-9 partial-icosahedron.

70-Ytterbium-174 needs a Li-9 partial-icosahedron**.**

74-W-183 needs a Be-9 partial-icosahedron.

76-Osmium-192 needs a He-9. Nuclear model building seems to get more difficult after Ti-48 but this addition completes Shell 6.

80-Mercury-201 needs a Be-9 partial-icosahedron**.**

<84-Polonium-210> needs a Be-9 partial-icosahedron.

<86-Radon-219> needs the equivalent of a He-9.partial-icosahedron.

<88-Radium-228> needs the equivalent of a He-9 partial-icosahedron and then rearrangement, maybe, to the **Key** configuration.

It turns out that among the element-isotopes ranked number 1, two were completed by adding He-9-type partial-icosahedrons; two with B-9-types, six with Li-9-types and one with Be-9-type.

6. An Icosahedron-Based, Periodic Table of the Elements

Einstein once said, more or less, that if two theories are proposed that explain observed phenomena, the one that is easiest to understand and accomplishes the objectives, is probably the better one. **Table III** is a little different than any I have ever seen, mostly because I determined to build icosahedron-based models as discussed. That places Group Ia and Ib elements in the same column this time, not as separate groups on each side of Group VIII. Noble gasses are found in Group VIII too, not in a separate Group. It doesn't separate the transition elements from the body of the table either. It includes some trans-fermium elements but fundamentally it reports what was found with the models and the isotope tables. **Table III** also suggests that every nuclide may have a possible geometrical structure based on individual nucleons or ultimately using atoms. The new, theoretical table based on the **Key** underlined element-isotopes is for use by theoretical scientists in the 21st century, to explain fundamentals in chemistry and physics in ways never done before. It also leaves many isotopes whose structures must be deduced or explained by future scientists. Hopefully they will work out the best arrangement of the completed table and maybe what structures the intermediate, not underlined element-isotopes have.

As usual it appears, it seems, it looks, it could be etc., that the completed icosahedron models are made primarily from hydrogen-3 and helium-3 nuclei through titanium-48 and then, maybe, possibly, imaginably, construction gets much more difficult. (My eyesight has gotten so bad that I can barely see what is on the page I have typed on a 17-inch monitor, so the details will be left to the classical physicists and chemists to work out properly.) It still looks, seems, appears, might be, that models can rearrange their charge distributions to get to their ground-state configurations.

TABLE III. A MODEL-BASED, PERIODIC TABLE OF THE ELEMENTS.

Group V	Group VI	Group VII	Group VIII	Group I	Group II	Group III	Group IV	Shell	Outer Electrons
		H	$_{2}He^{4}$	(H)					$(1s)^2$
				Li	Be	B	C^{12}	1	$(2s)^2(2p)^2$
N	O	F	Ne^{20}						$\mathbf{(2s)^2(2p)^6}$
				Na	Mg	Al	$_{14}Si^{30}$		$(3s)^2(3p)^2$
P	S	Cl	$_{18}Ar^{40}$						$(3s)^2(3p)^6$
				$_{19}K^{39}$	$_{20}Ca^{40}$	Sc 3d	$_{22}\mathbf{Ti}^{48}$	2	$(3d)^2(4s)^2$
V	Cr	Mn	$_{26}Fe^{57}$	Co	Ni				
				Cu	$_{30}Zn^{66}$	Ga	$_{32}Ge^{74}$		
$_{33}As^{75}$	Se	Br	$_{36}\mathbf{Kr}^{84}$					3	$(4s)^2(4p)^6$
				Rb	Sr	Y 4d	$_{40}Zr^{93}$		
$_{41}Nb^{93}$	Mo	Tc	$_{44}Ru^{102}$	Rh	Pd				
				Ag	$_{48}Cd^{111}$	In	$_{50}\mathbf{Sn}^{120}$	4	$(5s)^2(5p)^2$
Sb	Te	I	$_{54}Xe^{129}$						$\mathbf{(5s)^2(5p)^6}$
				Cs	$_{56}\mathbf{Ba}^{138}$	La 4f	Ce		
Pr	Nd	Pm	$_{62}Sm^{147}$	Eu	$_{64}\mathbf{Gd}^{156}$	Tb	Dy	5	$(5d)^1(6s)^2$

			$_{67}Ho^{165}$	Er					
				Tm	$_{70}Yb^{174}$	Lu 5d	Hf		
Ta	$_{74}W^{183}$	Re	**$_{76}Os^{192}$**	Ir	Pt			6	**$(5d)^6(6s)^2$**
				Au	$_{80}Hg^{201}$	Tl	Pb		
Bi	$_{84}Po^{210}$	At	$_{86}Rn^{219}$						**$(6s)^2(6p)^6$**
				Fr	**$_{88}Ra^{228}$**	Ac	Th	7	**$(6p)^6(7s)^2$**
Pa	$_{92}U^{237}$	Np^{237}	$_{94}Pu^{237}$	Am	$_{96}Cm^{246}$	Bk	Cf		
			Es	$_{100}$**Fm**					
				Md	**$_{102}No^{264}$**	$_{103}Lr$	$_{104}Rf$	8	
$_{105}Ha$	$_{106}Sg$	$_{107}Bh$	$_{108}Hs$	$_{109}Mt$	$_{110}Ds$				
				111	112	113	114		

Table III is almost self explanatory but building the models in the manner described may not be correct. Unfortunately you get the same relationships if you build out from the core carbon in four linear rays of icosahedrons joined at para-face-positions on icosahedrons. Maybe it should grow with spiral rays using pairs of tetrahedrally directed faces, just two at a time, from level L out to Q. Maybe the spiral helix of DNA is determined by the carbon, nitrogen, oxygen etc. nuclei linked together. What is presented is the most compact structure that appeared best, in my imagination, to make the nuclei small and compact and understandable with what is already known about nuclei.

It was very interesting to note that, after uranium, the next completed, icosahedron-based, model would be A = 237 and *Science News* reported Oct. 26, 2002, p. 259, neptunium-237 is the isotope with a critical mass between uranium-235 and plutonium-238 that is about as fissionable as U-235 and is the most common and most stable of neptunium's isotopes with $t_{1/2} = 2.14 \times 10^6$ yrs. There is no relatively abundant U-237 or Pu-237 so Np-237 is recognized as a very significant isotope. All seem good, but U-237 fits the appearance of the periodic table better so it was chosen.

Using more recent periodic tables including M. Jacoby's in *C&E News*, Mar 23, 1998, p. 48, added some additional elements. April 16, 1979 *C&E News* p. 49 showed Seaborg's idea of a longer periodic table and element 106 was shown as a Group VI element to go along with rutherfordium and hahnium in Groups IV and V. July 7, 1997 M. Jacoby, in *C&E News* p. 5, says seaborgium, $_{106}$Sg, is the heaviest Group VI element. I agree using the models found here **but** its longest, half-life isotope should probably be found at a higher mass number.

Some More Observations about Table III: 1. Possible, completed-shell elements arsenic-75 and niobium-93 are entered in Group V and were kept there, but it is suggested that zirconium and germanium found in group IV are better choices and their failure to fill the initial equation results must be charge distribution problems, possibly related to those found in unstable technetium where no number of neutrons will stabilize 43 protons and provide a stable isotope.

2. Completed icosahedron elements W-183, Po-210, U-237 and $_{106}$Sg are found in Group VI.
3. Two **Key** elements, **Kr-84** and **Os-192** are found in Group VIII along with chosen elements Fe-57, Ru-102, Xe-129, Sm-147, and Rn-219, the other completed-icosahedron models found there. With He-4, Ne-20 and Ar-40 that includes all the noble gas elements. In fact 12 of the 14 elements in group VIII are made into nuclei formed from regular polyhedrons.
4. K-39, with three stable isotopes, and fermium are the only completed-icosahedron-based models in Group I.
5. Three **Key** elements, **Gd-156** and **Ra-228** and **No-264** appear in Group II along with completed-icosahedron models Ca-40, Zn-66, Cd-111, Ba-138, Yb-174, Hg-261 and Cm-246. Amazing? Maybe it is just logical. Can the models be right? It seems they can be.

6. Three **completed-shell** elements appear in Group IV. **C-12**, **Ti-48** and **Sn-120**. Si-30 may be a carbon core with two fused Be-9's attached to complete two more icosahedrons. That provides (6 protons + 6 neutrons + 8 protons + 10 neutrons) = (14 protons + 16 neutrons) = 30 nucleons = Si-30 with just two attached, completed icosahedrons. It is so easy to make these models it is close to frightening.
7. As-75's neighboring element Ge-74 and Nb-93's neighboring element <Zr-93> fit in group IV. <Ge-75> has three, unstable, isomeric isotopes among stable germanium isotopes 70, 72, 73, 74 and 76. Unstable <Zr-93> would be ranked fourth among 5 zirconium isotopes; 90, 91, 92, 94 and 96, if its half-life of 1.5 x 10^6 years was called stable too. There is a lot still to be learned but it all seems to be making good sense.

Therefore the models themselves, and the systematic entry of the models into the *new,* periodic table, both speak to the probability of their existence, not just their "coincidental," statistical probability, but their actual existence as specific nucleon models.

Those interested in a very recent edition of a Periodic Table might like to look at *Chemical & Engineering News*, p.29, Sept. 8, 2003 to see the elements separated into yellow, green, blue, white and orange groups. That is what I would consider an antique-type, a chemist's arrangement, convenient to rounding out chemical properties. I hope the new Table III will develop a new, even better understanding of chemistry and physics including compound formation, bond angles and bonding structures. If scientists would just solve the fundamental problemss of their sciences they would understand them much better.

Table IV, coming up, shows where all the chosen element-isotopes would fall in the step-wise completion of atomic, electron shells. The remarkable order in their placement, in second position in many rows and also terminating almost every row is obvious and supports the models proposed here as having fixed structures with specific meanings relative to the location of electrons in the atoms. Some way, some time, it should be possible to explain the position of electrons in atoms and the positions of protons in their nuclei so that everyone will understand why each consecutive electron takes that one and only one position in the atom. I would like to say again that the nucleus determines the position of electrons in atoms and necessarily determines the chemistry as well. It seems there is room in and around the consecutive levels of icosahedrons to hold the levels of electrons and explain the Atomic Interpretation of the Periodic Table as in **Table IV**.

Table IV. A Different Look at the Arrangement of the Elements.

H							He				
Li	Be										
		B	**C**	N	O	F	**Ne**				
Na	Mg										
		Al	(Si)	P	S	Cl	(Ar)				
K	(Ca)										
		Sc	**Ti**	V	Cr	Mn	**Fe**	Co	Ni	Cu	**Zn**
		Ga	(Ge)	**As**	Se	Br	**Kr**				
Rb	Sr										
		Y	(Zr)	**Nb**	Mo	Tc	**Ru**	Rh	Pd	Ag	**Cd**
		In	**Sn**	Sb	Te	I	**Xe**				
Cs	**Ba**										
		La	Ce	Pr	Nd	Pm	**Sm**	Eu	**Gd**		
		Tb	Dy	**Ho**	Er	Tm	**Yb**				
		Lu	Hf	Ta	**W**	Re	**Os**	Ir	Pt	Au	**Hg**
		Tl	Pb	Bi	**Po**	At	**Rn**				
Fr	**Ra**										
		Ac	Th	Pa	**U**	Np	Pu	Am	Cm		
		Bk	Cf	Es	**Fm**	Md	No				
		Lr	Rf	Ha	Sg	Bh	Hs	Mt	Ds	111	112
		113	114								

All the bold, red, underlined elements are the **Key** elements. The black, underlined elements are the chosen, individually-completed, icosahedron-based elements. This is a periodic table arrangement of the elements according to their atomic numbers. (Ar), (Si), (Ca), (Ge) and (Zr) seem to be special- case elements closely related to structurally-completed, icosahedron-based elements that might be preferred, logically, to the actually more abundant, probably structurally-related, element-isotopes found adjacent to them in the table. In this way of looking at them argon and calcium are adjacent to potassium. But Finkelnburg on p. 130 (in 1964) has the electron-shell-derived configurations of 103 elements of the Periodic Table arranged according to the build-up principle. His arrangement has succeeding rows ending with the elements in the order shown:

: He, Be, Ne, Mg, Ar, Zn, Ca, Kr, Cd, Lu, Sr, Xe, Hg, (Lw = Lr), Ba, Rn and Ra.

Comparing this arrangement with Table IV (created here in 2004) leads us to a slightly different sequence that includes most of these elements or closely related ones.

He, Be, Ne, Mg, Ar, Ca, Zn, Kr, Sr, Cd, Xe, Ba, Gd, Yb, Hg, Rn, and Ra. Table IV has Gd and Yb in addition to the others, but Yb is adjacent to Lu and Lr is beyond the scope of what I chose to project into the periodic tables. Transition element Gd is simply shown where it is found in Table IV.

Pauling said he was looking for a relationship between specific, geometrical structures and the chemical elements. Looking again at Table IV's underlined, completed icosahedron elements, it was more apparent than ever that they met the standards for a surprisingly close relationship between two entirely different theories. He, Ne, Ar, Kr, Xe and Rn are followed by alkali and alkaline earth metals as neatly as you could want them. These completed geometric structures seem to have absorbed hydrogen atoms and can then lose the electrons easily.

7. A Long Form Table of the Elements.

Now one might ask if a simpler table can be created. **Table V** might help in understanding the elements because it is based on the models and electron Shells. It is related to Alfred Werner's long form of the periodic table which was arranged in the order of atomic number also way back before 1916. That arranges the elements in the order line 1 has 2; lines 2 and 3 each adds 8 for a total of 16; lines 4 and 5 adds 18; lines 6 and 7 adds 18; lines 8,9 and 10 together add 32; and lines 11 and 12 adds 28 more elements. Four more elements would seem to be needed to get to the next noble gas and complete the atomic electron spectroscopic sequence of 2, 8, 8, 18, 18, 32 and 28 plus 4.

Table V. Ground-State, Geometrical-Figures Interpretation of the Periodic Table.

Lines	Groups ➔	I	II	III	IV	V	VI	VII	VIII..........			.,Ia	IIa	IIIa	IVa
1.		H						H	**He**						
2.		Li	Be	B	C	N	**O**	F	**Ne**						
3.		Na	Mg	Al	**Si**	P	**S**	Cl	(Ar)						
4.		**K**	(Ca)	Sc	Ti	V	Cr	Mn	**Fe**	Co	Ni	Cu	**Zn**	Ga	(Ge)
5.						**As**	Se	Br	**Kr**						
6.		Rb	Sr	Y	(Zr)	**Nb**	Mo	Tc	**Ru**	Rh	Pd	Ag	**Cd**	In	**Sn**
7.						Sb	Te	I	**Xe**						
8.		Cs	**Ba**	La	Ce	Pr	Nd	Pm	**Sm**	Eu	**Gd**	Tb	(Dy)	**Ho**	(Er)
9.		Tm	**Yb**	Lu	Hf	Ta	**W**	Re	**Os**	Ir	Pt	Au	**Hg**	Tl	(Pb)
10.						Bi	**Po**	At	**Rn**						
11.		Fr	**Ra**	Ac	Th	Pa	**U**	Np	(**Pu**)	Am	**Cm**	Bk	(Cf)	**Es**	(Fm)
12.		Md	**No**	Lr	Rf	Ha	**Sg**	Bh	**Hs**	Mt	Ds	111	**112**	113	(114)
13.						115	**116**	117	**118**						

Red, not underlined, elements in Table V have the fundamental, geometrical structures involved in nucleon-atomic-model development based on helium, deuterium, tritium or hydrogen. **Red**, **bold- print, underlined** elements are the completed, icosahedron-based elements described as **Key** elements. Elements enclosed in parentheses are suggested to be isobars of adjacent element-isotopes of completed icosahedron-based elements and possibly being completed icosahedron-based nuclei themselves but maybe they have unique structures. **Bold, black, underlined** elements are the remaining, selected or *chosen*, individually-completed, icosahedron-based elements. It is hoped the arrangements presented will permit derivation of the changing rules required for construction of the models of elements in between the chosen elements.

Table V surprised me with its extremely orderly arrangements of the elements that finally ended up as shown. Why no one assembled the elements this way before was a disappointing thought until I read in John Emsley's book *The Elements* (1998) how the periodic tables gradually evolved to the mix it is today. As seen here it is a presentation in logic where elements made with completed, regular, polyhedron structures assume their distinctive positions, more or less, in conformance with what is known about the chemistry of the elements.

Line 1 shows the first geometric nuclear structure used to build nuclei, the one with the highest binding energy, a tetrahedron, the helium nucleus, and it was placed in Group VIII.

Line 2 has the **Key**, carbon-icosahedron nucleus essential to many subsequent nuclear constructions just as oxygen and neon can be built on icosahedral cores as described earlier.

Line 3 shows silicon, sulfur and argon. The first two very possibly are simple additions of nucleons to a carbon core, and the most abundant isotope of argon, with mass 40, may be the first lead to new insights into the chemistry of a poly-icosahedron-formed nucleus. It fits at the end of the line in Group VIII. I would still like to know if silicon-30, the third most abundant silicon isotope, can exist as an icosahedral carbon core with two fused icosahedral structures formed by the addition of two beryllium-9 partial-icosahedrons or if it is an icosahedron with sixteen added nucleons, eight protons and eight neutrons added to individual, appropriate faces to provide silicone-28, the most abundant silicone isotope. Silicon-30 could probably lose two neutrons easily to get to Si-28. Would one of those models or a mixture of those models be more prone to melt, fuse and form a glass? It is questions like this that delay my progress but with which I see them being answered by careful study of physical, nuclear models that will ultimately lead to finite answers to the mysteries of nuclear physics and nuclear chemistry.

Line 4 starts with **potassium**, includes the second **Key** element, **titanium** and ends with (germanium) an isobar of **arsenic**,

Line 5 begins with **arsenic**, which has only one stable isotope , and ends with **krypton**.

Line 6 ends with **tin,** and that element has more stable isotopes than any other.

Line 7 ends with **xenon.**

Line 8 puts a stress on **barium**, in Group II, and ends with (erbium), which is adjacent to **holmium-165** which, like arsenic, has only one stable isotope.

Williams, on p. 132, says magnesium-24 is a nuclear, doubly, magic nucleus with 12 neutrons and twelve protons in that, its most stable isotope; calcium-40 too is doubly magic with 20 neutrons and 20 protons in its most stable isotope; strontium-88's most stable isotope has 50 neutrons making it a mono, nuclear, magic number isotope. It does not seem to be surprising that **barium-138** has 82 neutrons making it too a mono, nuclear, magic number isotope in its most abundant, stable state. More and more evidence indicates that the models and physics and chemistry are coming together and the pairing of like nucleons can be found more easily. Magnesium is two elements after neon. Calcium is two elements after argon. Strontium two after krypton. Barium is two elements after xenon and radium two elements after radon. Those are fundamental relationships. Maybe the line-ending, pair of elements, **Ho** (Er), is a consequence of **gadolinium** completing **Shell 5** in its nucleus. Maybe with the start of the new **Shell 6**, noticeably farther from the core carbon-nucleus, that determines the new order of elements with fewer protons per **Key** interval, now down to 12 per 36 added nucleons.

Line 9 ends with (Pb) the second element after **Hg**, in much the same way (germanium) is two elements after **Zn** in Line 4 and **Sn** is two elements after **Cd** in Line 6.

Line 10 ends in the radioactive gas **Rn,** as expected and where expected.

Lines 11, 12 and 13 are added without any firm conviction about which elements are completed polyhedrons, only that they are arranged in a manner which correlates with atomic intervals in Lines 8, 9 and 10 and seem possible with the knowledge that Fr, **Ra**, Ac, Th , Pa and **U** fit perfectly with earlier parts of Table V. This preliminary conclusion about a loss of more logical, model, identity positions of transuranium elements was reinforced by the report that neptunium is more like uranium than like rhenium; that the longer, half-life, transuranium elements actually occur at plutonium when compared to neptunium; **curium** compared to americium, californium compared to berkelium but einsteinium compared to **fermium**, The most noticeable discrepancy is how fast the elements develop at a sequence of lower increases in mass number.

Np-237 has a longer half-life than U-238 when comparing elements that way. Am-243 has lower mass than Pu-244 and a longer half-life. Berkelium and Curium have the same mass weights so they are isobars but berkelium's half-life is about 500 years longer than curium's. Growth of transuranium elements seems quite different from the rest of the periodic table. But maybe, if the right isotopes were formed according to the construction of completed icosahedrons more stable half-life isotopes would be found. It seems it is not easy trying to make new specific isotopes.

Line 13 was added to provide a total of 32 elements after **radon** in order to conform with atomic, shell-level predictions, if not reality.

This brings us to looking at the individual *Groups in Table V.*.

Group I has only one chosen element, potassium, just maybe because it is the first, chosen, multi-icosahedron-represented element and the nucleon numbers 12 + (3 x 9 = 27) = 39 works out so easily to provide mass number 39, its most abundant (93%), stable, atomic mass isotope. But it is noted that argon-40 and nuclear, doubly, magic number isotope calcium-40 are each more than 99% abundance among their stable isotopes. It is easy to see how they may be related to potassium and simply part of constructing the proposed isotope models systematically.

Group II hosts completed icosahedron structures for possibly (calcium), **barium**, **ytterbium**, **radium** and **nobelium.**

Group III shows no completed polyhedron structures.

Group IV has **Key** elements **carbon** and **titanium** and possibly completed icosahedron isotopes at (Si) and (Zr).

Group V shows only **arsenic** and **niobium**. All this implies to me that nucleon positions can be depicted in unique positions in ground-state, element-isotope models.

Group VI seems to develop a new way to build structures beginning with Shell 6, at terbium, that provides completed icosahedrons at **tungsten**, **polonium**, **uranium** and **seaborgium**.

Group VII contains no obvious, completed, icosahedron-based models, probably because so many models are completed in the next Group.

Group VIII shows (argon) as a possible one, **iron**, **krypton**, **ruthenium**. **xenon**, **samarium**, **osmium** and **radon**,, and possibly (**Pu**) and **Hs**. Helium and neon were included in Group VIII because of their lack of chemistry and the long history developed for these elements. The other completed icosahedrons in the extended-Group VIII elements are **gadolinium** and **curium.** That means Group VIII all together has a surprising total of 12 completed models

Group Ia has no special representatives.

Group IIa on the other hand has **zinc**, **cadmium**, possibly_dysprosium, **mercury** and possibly (californium).

Group IIIa has **holmium** and possibly **einsteinium** if the proposed evolution for their formation is correct.

Group IVa may have (germanium) along with **tin**, maybe (erbium), (lead) and (fermium). I was stationed in Oberramstat bei Darmstat during the second world war and was surprised to find that Darmstat, the last named element included in Table V, got an element named after it. But I hope they find a use for the models developed in this paper to find a way to build new elements or isotopes that will have longer half-lives because of improved stability.

Needless to say I was disappointed that the **Key** elements didn't all fall in some perfectly orderly fashion but I have to admit they earned a place in the way I developed Table V. The arrangement confirms an entirely new look at the formation of nuclei and atoms and the difficulties that are encountered and will have to be explained. As the number of neutrons increase in the chosen elements the arrangement in Table V changed too. Counting the number of elements in Lines leading up to and including **Key** elements gives 2 + 4 to Carbon; 4 + 8 + 4 to **titanium**; 10 + 4 to **Kr**; 14 to **Sn**; 4 + 10 to **Gd**; 4 + 8 to **Os**; 6 + 4 + 2 to **Ra** and 12 to (Fm) . This confirms the sequence 6, 16, 14, 14, 14, 12, 12 and 12 shown in Table II. I am sorry to have created another periodic table that seems to be arranged only in the order of their atomic number but I think it well help someone or many people find the fundamental relationships between the models, atomic isotope Shells and nuclear Shells, thereby uniting nuclear chemistry and physics. At the same time I am amazed that Alfred Werner could have come up with a long form of the Periodic Table at the beginning of the twentieth century and so little progress seems to have been made since then in understanding the structure of nuclei and atoms.

Considerations Related to Atomic Fission.

Looking back in time at nuclear fission as it was presented at Argonne National Laboratories in Chicago in the late 1950's using the *Source book on Atomic Energy* by S. Glasstone (1958) the primary fission products of thermal-neutron-induced fission of **U**-235 ranged from atomic mass numbers 72 to 160 "believed to be isotopes of zinc to gadolinium." (pp. 394-395) Most of the fission products fell into two groups; a "light" group with atomic mass units 85 to 104 and a "heavy" group with A = 130 to 149. These would include the stable nuclei krypton-86 to ruthenium-104 as the limits in the light group and stable nuclei barium-130 to samarium-149 in the heavy group found at these limits. Along with this, the most probable products of the fission have atomic masses of 95 and 139 which would be very close to niobium and barium. All of these are isotopes of **Key** elements shown in Table I.. All six are icosahedrally-complete, selected elements found in **Tables I** to **III**. This is an entirely different kind of evidence that these new models can be real and are related to known chemical facts. Completed-icosahedron models would be less abundant than those with broken icosahedrons on their peripheries occurring as a result of atomic fission, or maybe as a result of the **Big Bang**, But U-238 plus a neutron gives U-239 which typically breaks down to "**Ba**-138 and Tc-101." Technetium is important to the Tables because it is an isobar of Ruthenium-101, a stable, one-less-neutron-neighbor, of **Ru**-102, another one of the *selected* elements. And one more and one less nucleon are common occurrences among atomic nuclei and atoms to say the least..The very old problem of how to explain the unsymmetrical nature of the fission process may be fundamentally related to the icosahedron-based structures of the nuclei and can be explained looking at the models that are created with the fused-icosahedrons concept and how they might be torn apart. It was taught that if the U-235 nucleus were torn apart to give equal halves, the mass of each would be 117-118, which could be Sn-117 or Sn-118, stable isotopes of **Sn-120**, The fact that this is only about 0.01% of the fission products is probably because the single, core-carbon-icosahedron is protected by 25 other icosahedrons that can capture thermal neutrons and it appears titanium-like, nuclear structures in the U-235 nucleus can capture neutrons and result in fission too.

Observations:

1. Models for selected isotope nuclei of specially chosen atoms of elements in the periodic table of the elements have been suggested for consideration by nuclear chemists, physicists, radiologists and the finest aspiring scientists of the 21st century. It may be a miracle if the suggested structures actually represent fundamental relationships to chemical atoms but all it takes to build with them is to reserve from ones mind some of the current pipe-dreams of nuclear science like:

1. The *religion* of quantum-mechanics, that no one can understand, may now be interpreted.

2. Nuclear structure has no influence on the structure of the atom except its central charge and the effects on the conglomerate cloud of electrons around it.

3. The nuclear charge is not directed.

What is presented here, I believe, is evidence toward models of ground state nuclei whose structure will be related to:

1. Nuclear structures that determine the chemistry of atoms.
2. Nuclear structures that determine the distribution of the electrons in atoms.
3. Carbon atom which has the shape of an icosahedron formed from two helium-3 and two hydrogen-3 atoms, which equates to 12 nucleons with two electrons on the inside and four valence electrons on the outside of the nucleus. Fascinating is the observation that carbon icosahedron could occur in right and left hand forms which might lead to amino acids occurring in right and left hand forms and someday may explain why amino acids in living organisms occur preferentially in the left hand form.
4. The next four selected isotopes are formed by fusing beryllium-9 nuclei to tetrahedrally-directed faces of the carbon core.
5. But partial icosahedrons of lithium-9 nuclei appear to become the added species to make some higher-atomic-weight, selected, nuclear isotopes of the elements. In fact, the equivalent of four are needed to get from gadolinium-156 to osmium-192 to maintain the right ratio of neutrons to protons..
6. Chemist's atoms usually would be built using neutrons, hydrogen atoms, hydrogen-2 atoms, hydrogen-3 atoms, lithium-9 atoms, beryllium-9 atoms and even carbon-9 atoms as it is described here.
7. Many physicist's nuclei would be built using neutrons, protons, deuterons, tritons. (fully ionized tritium atoms) etc., as needed, but now they can have structures to use as targets or to understand the physics better.
8. Maybe these models will get chemistry and physics off to a better start in the 21st century than has been achieved in the past. At the very least it would appear that there is a new numerical relationship between elements in the periodic table which identifies many of what seem to be preferred, stable element-isotopes around which to build isotopes and missing elements.
9. It appears someone should be attempting to fuse four potassium-39 atoms together to obtain **gadolinium-156**, because 4 x 39 = 156. Neutral atoms may not be too difficult to compress and cause fusion to occur. Anyone can dream too.

In going back to Williams' book *Nuclear and Particle Physics*, p. 1, I rediscovered atomic physics is said to be the physics of the electronic structure of atoms which is not markedly affected by the properties of the nucleus, apart from charge, spin and magnetic moments. "The behavior of the nucleus is little affected by the electronic structure surrounding it." so nuclear physics separates the cloud of electrons from the tiny neutron-proton nuclear mass and has missed what nuclear physics and atomic physics really have in common.

Subsequent Observations:

1. The chemist's elements are atoms and nuclei composed of a broader sense of nucleons, that is of neutrons, protons and electrons, that can be arranged to provide the chemistry as needed..

2. Carbon is an icosahedron made from two hydrogen-3 atoms and two helium-3 atoms touching one another on the surface of the icosahedron. The usual structure would have four electrons arranged above those atoms. Two of the helium atom outer electrons would have opposing electrons inside the atom. But I believe that those two inner electrons could very well interact with two 1s helium protons forming equivalent-neutrons forming carbon nucleus with an obvious valence of just four. The four, equivalent hydrogen-3 faces are left with identical properties and that is the chemists hybridized carbon-13 atomic nucleus. Therefore the chemistry of carbon is being elucidated gradually and its spectroscopic description is being used effectively to identify all the nucleons.

3. If one makes nitrogen-14 to resemble the chemist's and physicist's best known structure one would start with carbon and add one hydrogen atom placed arbitrarily on the n-n-p-hydrogen-3-face *chi* opposite the p-p-n-helium-3-face *bch.* which would receive an added neutron forming a prolate or football-shaped nitrogen nucleus with two, more or less, kinds of tetrahedral-shaped helium-4's in para position to each other. But these positions will explain the valences of nitrogen. The addition of the neutron encloses a valence electron inside the helium-4-type moiety formed reducing the valence of the carbon core to three as it blocks access to the carbon core at that position.. Its other electron will be inside the carbon core helium-3 face on the other side of where it became attached. The proton addition, to a hydrogen-3 face in para-position on the carbon core, provides a differently-arranged helium-4-type moiety. This seems to provide an explanation for what is called an *unshared pair* of electrons on nitrogen atoms. From what has been developed here, it easily provides one exposed electron at least. An exposed electron on one of those helium-4 faces would be different than the three tetrahedrally-positioned ones exposed on carbon as far as electrostatic forces are concerned. It should account for the behavior of nitrogen atom to form covalent compounds like ammonia using those tetrahedrally-positioned faces to this unique helium-4 position, and for ammonia's ability to form ammonium ion . The *chi* and *bch* faces were chosen because they opposed one another on the core. I do not know if it was the best choice among several available but it certainly worked out well. Another surprising observation might be that these two helium-type moieties may be fixed in those positions and not helium-like atoms at all.

Maybe one can consider that the carbon nucleus has 0 spin, which it has. and a proton and a neutron on opposites sides of the carbon core can be easily imagined to be unpaired and that pair of nucleons would corresponds to deuterium that has a spin of 1. Therefore nitrogen nucleus should have a spin of

1 and it does. The prolate shape fits what is needed in chemistry too. When the ammonia model was imagined with three hydrogen atoms bonded to the nitrogen nucleus, it was easy to imagine it as one of the nitrogen inversion models. Considering the mass of the nucleus relative to the mass of three protons, I chose to have the nitrogen atom invert all its protons to neutrons and all its neutrons to protons as electrons moved back and forth within the nucleus or on its surface and as hydrogens moved back and forth above the nitrogen core to be located in the proper bonding areas of the nitrogen models. **Fig. 10** shows what could be representations of one nitrogen atom in two inversion resonance appearances.

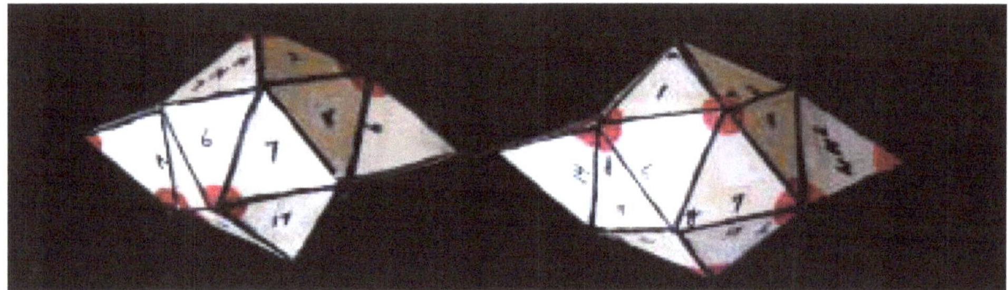

Fig. 10. Nitrogen Inversion Models.

Now nuclear chemistry seems to be related very well to chemistry itself as one thinks of stereoisomers as inversion structures. When you have physical models to study, solving problems seem to be much easier than wave equations and "renormalizations" might be. Anyway that is why the nucleus was chosen in the football shape as opposed to making the additions at tetrahedral locations on the carbon core.

This could be significant because unstable *carbon-14* nucleus loses a beta particle to provide stable nitrogen-14. I would normally add two neutrons to a carbon core model on two helium-3 faces, as was done building most of the chosen poly-icosahedron models using tetrahedrally directed faces. Losing a beta particle from either added neutron would produce a *lithium-4* tetrahedron at that location. The isotope table shows no known *lithium-4* atom but maybe it could exist as part of an unstable carbon atom. If one starts with a carbon-14 model and a β-particle is lost in any way whatever from some neutron on the atom, then you make a new structure for nitrogen-14 isotope. It is hoped someone has checked to see if nitrogen from radioactive *C-14* is the same as known normal N-14 that would work in inversion models. Of course maybe the new-type of N-14 would rearrange to give old N-14. But I think nuclei have fixed nucleon positions and only charges move around according to rules still to be learned if my propositions get that far. Now what would oxygen-16 look like? That means oxygen has four valence faces, two slightly different from the other two if they are forced to maintain their character as shown by the model.

Take standard carbon-12 model and add two neutrons, one each to faces 5 and 17. Then add two protons, one each to faces 10 and 20. This provides oxygen-16 made from four helium-4 atoms joined at corners that leave four tetrahedrally-directed hydrogen-3 faces, 2 and 15, and two helium-3 faces,12 and 18. It appears oxygen has two helium-4 structures that could represent what is needed to provide two electron pairs. Faces 12 and 18 would represent covalent valence-electron faces. The electron pairs associated with the second level of protons could repel the valence-electrons away from each other slightly to provide the 116° bond angle in ozone. Resonance double bond character in ozone would be provided by faces 2 and 15 to reduce the O-O bond distance to 1.27 A. HOOH has an O-O distance of 1.49 A. Making the double bond connection too could help pull the O-O-O angle greater than the tetrahedral angle 109°.

Just maybe in water, with hydrogens bonded to faces 12 and 16, the hydrogens are small enough to be attracted by the unshared pairs and decrease the bond angle to 106°. The other two tetrahedrally directed, hydrogen-3 faces 2 and 15, would presumably be involved one at a tome in the formation of oxonium ions and four coordinate compounds. (Cotton and Wilkinson, p. 275).

But going back to nuclei and adding one neutron at a time to faces 12 and 18 could provide stable isotopes oxygen-17 and -18. Then adding a proton to face 2 *or* face 15 would provide fluorine-19, the only stable isotope of that element. Adding one more proton to one of those two fluorine nuclei would provide neon 20. Is there only one fluorine-20? Maybe oxygen nucleons should be arranged differently. It looks like neon-20 is made from eight alpha-particles with their bases joined together. That nucleons on the second level look like tetrahedrons and the second level nucleons all oppose each other on opposite sides of the icosahedron. The configuration and arrangement described here and earlier may explain the inertness of neon and lack of chemistry through its structural relationship to helium.

Everyone should know however that the more heads are working on a problem the better the ideas will be that come out of the solutions proposed. It is time to get everyone interested in this new approach to nuclear chemistry and physics so better answers are obtained..

Final Conclusions:

I. My primary aim in this paper is to make public the discovery that *"models of reality"* of the atoms and nuclei of chemistry and physics appear possible that may help explain the various valences of individual elements, the differences between isotopes of the same element, the numbers of isotopes of individual elements, nuclear and classical chemistry as we should learn it; complete explanations of nuclear physics as it has been experimentally developed, and maybe even someday an explanation of why wave mechanics can be applied to the motions of nucleons in nuclei and electrons in atoms.

What has been put together here must be more than a mere coincidence. I can no longer read the books that helped me get this far, because of failing eye sight that loses visual acuity soon after I begin reading, but I hope that many scientists will come to appreciate icosahedron-based nuclear models and be able to solve the problems presented here and still not answered in the course of working out the new structures. Certainly it will require some excellent electrostatic engineering specialists and excellent geometricians to relate physical model structures to crystallization structures, atomic orbitals to electrostatic fields and quantum mechanics properly so this introduction to *models of reality* can be confirmed and completed successfully. And so I propose this new beginning in nuclear physics and chemistry.

J. Alden Erikson

www.ingramcontent.com/pod-product-compliance
Lightning Source LLC
Chambersburg PA
CBHW051100180526
45172CB00002B/720
9781420852509